SHORTS 330 and 360

P. R. SMITH

Copyright © Jane's Publishing Company Limited 1986
First published in the United Kingdom in 1986 by
Jane's Publishing Company Limited
238 City Road, London EC1V 2PU
in conjunction with DPR Marketing and Sales
37 Heath Road, Twickenham, Middlesex TW1 4AW

ISBN 0 7106 0425 4

Printed in the United Kingdom by Netherwood Dalton & Co Ltd

JANE'S

Cover illustrations

Front: **Air UK** (UK)
Air UK is a major British independent airline operating to many UK
airports as well as to cities in Belgium, Denmark, France, Holland and
Norway. The carrier was formed in 1980 through a merger of British
Island Airways, Air Anglia, Air Wales and Air Westward. The red,
white and blue 330s and 360s have the distinction of flying on more
international routes than those of any other Shorts operator worldwide. In July 1984 Air UK passed a milestone in aviation history
when it became the first British operator to fly an all-female crew on
one of its flights, which coincidentally employed a 360.
(Short Brothers Plc)

Rear: **Allegheny Commuter System** (AL)
The Allegheny Commuter System is part of the large American airline
US Air. The company is formed of different airlines which operate
flights on its behalf under the designation "AL". Airlines that make
up the ACS include Air Kentucky Airlines, Chautauqa Airlines,
Crown Airways, Fischer Brothers Aviation, Pennsylvania Airlines,
Pocono Airlines, Southern Jersey Airways and Suburban Airlines.
(Short Brothers Plc)

Right: **Imperial Airways** (II)
Until the company ceased operations in 1986, Imperial was a fast-
growing regional commuter carrier operating daily scheduled flights
to points in Southern California. The privately owned airline began
operations in 1967 and until flights ended flew to Bakersfield,
Carlsbad, Los Angeles, Ontario, Orange County, Palm Springs, San
Diego, San Louis, Obispa and Santa Barbara. *(Short Brothers Plc)*

Introduction

Initially known as the SD3-30, the Shorts 330 made its maiden flight as G-BSBH on 22 August 1974. The 30-seat aircraft is derived from the Skyvan, keeping many of the latter type's well proven characteristics, and is designed as a transport for commuter and regional operators. The 330 is a high-wing, strut-braced monoplane of a light alloy construction. In addition, a square-section fuselage gives an internal cabin width and height of 1.93 m (6 ft 4 in). The tail unit is a cantilever structure, comprising a fixed-incidence tailplane and full-span elevator with twin fin and rudders. Unlike the Skyvan, the 330 is equipped with a retractable undercarriage. The main wheels are located on short sponsons into which they retract hydraulically.

The first order for the 330 came from Command Airways of New York, and was for three aircraft which went into initial service with Time Air of Alberta on 24 August 1976. There were three versions of this aircraft available in 1986, including an all-freight version known as the Sherpa with a rear-loading ramp door. The first twenty-six 330s were powered by Pratt and Whitney Canada PT6A-45A turboprop engines, the next forty had PT6A-45Bs, and all aircraft after this are powered by PT6A-45R engines. In 1982 a military version, the 330-UTT, was announced. This example is able to accommodate 33 troops or 15 stretchers and five attendants. The United States Air Force ordered 18 Sherpas in 1984 and redesignated these aircraft C-23s. A maritime patrol variant was announced but was never produced. Had it been, it would have been known as the SD3-MR Seeker.

The Shorts 360 is a development of the 330, seating 36 passengers in a lengthened fuselage and having a new tail with single fin, plus many other updated features. Prior to the 360, Shorts designed two other variants but these were "in-house" productions and were never officially announced. The Shorts 333 would have carried 33 passengers and the Shorts 335 was a 35-seat variant; both designs were based on that of the 360, with a single tail-unit and retractable undercarriage. Designed for use as a short-haul aircraft over stage lengths of 222 km (138 miles), the 360 first flew on 1 June 1981, entering service with Suburban Airlines of Reading, Pennsylvania on 1 December 1982. The prototype, G-ROOM, was powered by PT6A-45 engines, but production aircraft use two PT6A-65R turboprops.

The Shorts 330 and 360 are quieter than most turboprops. This is due to the Pratt and Whitney engines which have, amongst other features, an efficient reduction gearbox permitting the propellers to revolve slowly and therefore more quietly. The cabins of both aircraft, which were designed by the specialists responsible for Boeing interiors, offer comfortable big-airliner seating and decor. Pressurization was not thought to be justifiable over the short stages flown, since this would have meant radical changes to the square section fuselage. Complement for both aircraft is two cockpit crew and one cabin attendant, although a few airlines have on occasions used two.

I would like to extend my sincere thanks to the people who were kind enough to provide transparencies for use in this book. They include Andy Clancey, Udo Schaefer and Short Brothers Plc. I am extremely grateful to Brian Richards and Tony Carder for their overall help.

Finally, I would like to dedicate this book to my mother, whose help and understanding in its preparation has been greatly appreciated.

TABLE OF COMPARISONS		
	330	**360**
Max. accommodation	33	36
Wing span	22.76 m (74 ft 8 in)	22.81 m (74 ft 10 in)
Length	17.69 m (58 ft 0.5 in)	21.59 m (70 ft 10 in)
Height	4.95 m (16 ft 3 in)	7.21 m (23 ft 8 in)
Max. t/o weight	10,387 kg (22 900 lb)	11,999 kg (26 453 lb)
Max. cruis. speed	352 km/h (218 mph)	393 km/h (244 mph)
Maximum range	1695 km (1053 miles)	1697 km (1055 miles)
Service ceiling	3500 m (11 500 ft)	3500 m (11 500 ft)

Opposite: **Manx Airlines** (JE)
Manx Airlines is a fast-growing British regional carrier, operating scheduled passenger flights to the Isle of Man and to points in England, Scotland and the Channel Islands, Northern Ireland and also to the Republic of Ireland. The airline has a flight base at Isle of Man Ronaldsway Airport, where it employs nearly one hundred staff. Manx was established in 1982 by British Midland Airways (which has a seventy-five per cent shareholding) and the British & Commonwealth Shipping Group. Typical routes served are Isle of Man-Liverpool-London (Heathrow), Isle of Man-Blackpool and Isle of Man-Belfast Harbour. Shorts 360 G-DASI is seen here just prior to take off from Belfast Harbour Airport. *(Short Brothers Plc)*

Opposite: **Aer Lingus Commuter** (EI)

Aer Lingus Commuter was formed in 1984 to cater for business and holiday traffic. A Shorts 330 (with the registration EI-BEG, although this was changed within 24 hours to EI-BEH) was used from their Dublin base to Liverpool and Leeds/Bradford, back in 1983. With the introduction of new services to Bristol, Cork, East Midlands, Edinburgh & Shannon, together with increased traffic on the other routes, it was announced that the 330 was in fact too small for Aer Lingus' needs, so an order was placed for four 360s, which gave the company the extra capacity needed. Aer Lingus Commuter's aircraft all seat 35 passengers and at one time used two cabin attendants instead of the standard one. With the 360s the company was able to increase its round trips on the more popular route, of Dublin to Liverpool, from four to five flights per week to twice daily. *(Andy Clancey Collection)*

Above: **Aeronaves Del Centro** (AG)

Aeronaves Del Centro is a Venezuelan domestic operator which commenced services on 7 October 1980. The decision to purchase the 330 was made on the basis that the Shorts type was the most economical aircraft to operate from some of the country's hot, high airports, as well as being able to take-off and land on unpaved strips. ADC's inaugural flight was between Maracay and Caracas with an additional service between Puerto Cabalio and Caracas. Today, the company is operating with increasing frequency around the north of Venezuela, flying to towns such as Barcelona, Barquesimeto, Maracaibo, Margarita, Puerto Ordaz and Valle de la Pasqua. *(Short Brothers Plc)*

Air Business (8A)

Air Business of Denmark, a wholly-owned subsidiary of the major carrier Maersk Air, was the first Scandinavian airline to adopt the Shorts 360 when it placed an order for two during December 1983. Since then the Esbjerg-based operator has added a third 360 to its fleet and all the aircraft are engaged in a busy programme of scheduled and charter domestic services, as well as international flights between the two oil centres of Esbjerg and Stavan-ger, with a link to Aarhus in Denmark. Air Business was in fact the first privately-owned Danish airline to be granted traffic rights for international scheduled flights and further applications have been filed. These include a route linking Billund and Aarhus with Oslo. In August 1985 the company completed extensive hangar and workshop facilities at its base, and all 360 maintenance is now carried out there by the airline's own staff.
(Short Brothers Plc)

Air Ecosse (SM)

One of the most eye-catching liveries worn by any Shorts aircraft must be that of the Datapost 360s, operated by Air Ecrosse. The company began services from Dyce Airport, Aberdeen, in 1977 and built up an extensive passenger and cargo network, pioneering the Royal Mail Datapost service. This involved the stripping out of all seats at the end of the passenger-carrying day and flying mail by night. The company serves Belfast, Dublin, Edin- burgh, Glasgow, Manchester, Isle of Man, Prest- wick and Wick from its Aberdeen base. Its Prest- wick services provide a feeder service for Air Canada and Northwest Orient's North American routes. Until recently, Air Ecosse served London Heathrow from Carlisle and Dundee, a link started on 4 April 1983, when it became the first Scottish commuter airline to serve the world's busiest inter- national airport. *(Short Brothers Plc)*

Opposite: **American Eagle** (AA)
American Eagle is a commuter air service system which has been developed by American Airlines to provide connecting feeder flights at major hub airports in conjunction with independent regional and commuter carriers. Services were inaugurated on 1 November 1984, when Metro Airlines commenced operating its established Dallas schedules under the American Eagle name. A month later flights of Chaparral Airlines also assumed the American Eagle identity. Shorts 360 aircraft were introduced into the Eagle colour scheme in 1985, when Simmons Airlines announced that it had been selected as the American Eagle carrier to serve Chicago's O'Hare International Airport and that from 1 October of that year, the Michigan operator would add eleven cities to the "AA" network. All the Simmons aircraft fly in the American Eagle livery. This example, N360MQ, is seen here on a typical flight. *(Short Brothers Plc)*

Above: **Atlanta Express** (FX)
Atlanta Express was formed in 1982 as a regional commuter airline operating Shorts 330 aircraft. The company had a base at Atlanta Airport and its route system, which covered a 200-mile radius, included flights to Hickory and Charlotte, North Carolina. Due to financial difficulties the company ceased operation in 1983. *(Andy Clancey Collection)*

9

Atlantic Southeast Airlines (EV)

Atlantic Southeast Airlines, the publicly-owned company, began operations in June 1979 with Shorts 360 aircraft. Following a merger four years later, the carrier absorbed Southeastern Airlines, another commuter carrier. This was the start of the company's rapid growth, as ASA today provides services to more than 20 points in Alabama, Florida, Georgia, Mississippi, North Carolina, South Carolina and Tennessee. Under a marketing agreement with Delta Air Lines, ASA was in 1986 operating as part of the "Delta Connection", which coordinates flights with Delta at their Atlanta and Memphis air hubs. The carrier retains its own identity, but like the other Connection members, use the Delta Air Lines flight code DL. *(N Raith)*

Avair — Irish Independent Airlines (KZ)

Avair announced that from the end of May 1983 it was to extend its route system to link Dublin with Blackpool, the Channel Islands and East Midlands. The addition of these new routes continued the expansion begun two years previously when the company, then engaged exclusively on charter work, launched its first scheduled operation using a Beech 99 on a twice-daily service between Dublin and Londonderry. Within five months the airline had acquired a Shorts 330 and expanded its operations with a three-times-daily service between Dublin and Belfast, later adding Cork, Ronaldsway on the Isle of Man and Waterford to its route system. Up to the period of 31 March 1983, the airline carried 50 000 passengers — an increase of more than 500 per cent over the previous year's figures. The appearance of Avair's colourful livery on the 330 was unfortunately short-lived, since having lost applications on routes and with traffic dropping, the company ceased operations on 23 February 1984 and the receivers were called in. *(Short Brothers Plc)*

Above: **British Midland Airways** (BD)
British Midland Airways was formed in 1938 as Derby Aviation, initially operating as an air training school. Having changed its name to Derby Airways, the company commenced charter flights in 1948 and five years later inaugurated scheduled operations between Wolverhampton and Jersey. The carrier adopted its present name in 1964. BMA is a fast-growing privately-owned airline operating scheduled and charter domestic and international flights, as well as offering an aircraft leasing service. The company has a seventy-five per cent shareholding in Manx Airlines, the regional Isle of Man carrier, as well as the same percentage in Loganair, the Scottish regional carrier. Some 14 regional UK airports are used to serve a variety of destinations, including Amsterdam and Brussels. The airline has a fleet of Douglas DC-9s, Shorts 360s and Fokker F-27s, with an order for the brand new British Aerospace ATP to replace the latter aircraft. The Shorts 360s are used on the company's East Midlands and Birmingham to London (Heathrow) routeings. G-BMHX is seen here sporting BMA's new livery. *(P Hornfeck)*

Opposite: **Brockway Air** (SS)
Brockway Air was formed in 1984 as a subsidiary of the Pennsylvania-based Brockway Corporation, who also own Crown Airways, the Allegheny Commuter company. In 1986 the airline had two other companies within its structure. Clinton Aero, which began operations under the parent company's name on 1 May 1984, and Air North, who joined four months later. The former had been serving points in central and upstate New York, the New York City metropolitan area and in Vermont, while the latter had a network from central New York and Vermont to the New York City, Boston, Philadelphia and Washington areas. Air North was formed in 1956 as Northern Airways, and began scheduled passenger flights eleven years later. The name Air North was not adopted until 1970. A Shorts 330 is seen here bearing the livery of the airline. *(Short Brothers Plc)*

Brymon Airways (BC)

Plymouth-based UK operator Brymon Airways, a member of the British Caledonian Commuter Service, used a Shorts 330 to provide a "new look", faster service from Birmingham to London (Gatwick). In addition to offering high standards of comfort and service, the 30-seat airliner reduced travel time on the route to 45 minutes. The airline commenced services in 1972 under the name Brymon Aviation, with a name change to Brymon Airways a year later. The company operates scheduled passenger services to other destinations, including Newquay, London (Heathrow), Isles of Scilly, Exeter, Jersey and Guernsey.
(R Vandervord)

CAAC — Civil Aviation Administration of China (CA)

A colourful ceremony held at Wuhan, Hubei Province, on 27 August 1985 marked the introduction of Shorts 360s to CAAC's network in the Guangzhou Region. The four aircraft involved brought the airline's fleet of this type to eight, completing an order placed in February 1985. The first four had been in service since that July, operating in the Shanghai Region. The introduction of the Shorts 360s was a major step in updating and supplementing CAAC's short-haul transport fleet of Antonov An-24 and Ilyushin Il-14 aircraft. The increasing demand for air services was being particularly concentrated around the four southern Special Economic Zones and fourteen coastal cities opened to foreign investment. Tourist destinations for the mass market had been limited to the principal cities like Beijing, Shanghai and Guangzhou (Canton). However, with the addition of the 360s to the fleet, the country was expected to open up to the tourist influx. *(Short Brothers Plc)*

Casair (ZR)

Casair was initially formed as an air taxi/charter company in February 1982, offering oil-related charters in Scotland and air taxi services throughout Europe. On 1 November 1980 the company took over Air UK's Teeside-Glasgow route, operating this on a three-times-daily basis utilising Cessnas. Shorts 330s, one of which, G-BJUK, is seen here, operated a weekday Teeside-London (Gatwick) service and seasonal weekend services to the Isle of Man and Guernsey from Teeside. The controlling interest was purchased by General Relays in 1982 and Casair's commuter operations were integrated with those of Eastern Airways of Humberside and Genair of Liverpool to form a new Genair. (Short Brothers Plc)

Comair Inc (OH)

The Cincinnati-based commuter airline Comair began services in March 1977, using a Piper Navajo aircraft, over the Cincinnati-Cleveland route the company had initially obtained. Today the carrier is a rapidly expanding regional airline that maintains scheduled passenger flights to over 20 destinations in the states of Ohio, Indiana, Michigan, Kentucky, Virginia, West Virginia, Tennessee, Wisconsin and Florida, as well as to the Canadian province of Ontario. In 1982 it was decided to purchase the Shorts 330; this was due to the increase in passenger traffic and the lack of capacity offered by the Embraer Bandeirante. In September 1984 Comair joined the new "Delta Connection" commuter network organized by Delta Air Lines, which provided flight coordination and passenger feed in Cincinnati. The company retains its independent operating identity as part of the "Delta Connection". *(Andy Clancey Collection)*

Opposite: **Command Airways** (DD)
Command Airways was the first company to operate the 330. Since the introduction of the type into service with the company, the aircraft have had a despatch reliability of over 99 per cent. The 330s were in 1986 flying an average thirteen-hour day, or 200 hours per month per aircraft. Before the introduction of the Shorts aircraft, Command operated a mixed fleet. However, after six months the company's Beech 99 was sold and a year later the Twin Otters were removed from the fleet, so standardizing on the Shorts aircraft. The company began services on 1 July 1966, when it commenced operations on the Poughkeepsie-New York (JFK) route. Nowadays Command operates a wide network of routes covering points in the Northeast states of New York, Massachusetts and New Hampshire. *(Short Brothers Plc)*

Below: **Coral Air** (VY)
Although no longer operating Shorts aircraft, Coral Air took delivery of a single example in 1981. The company provides scheduled passenger services within the US Virgin Islands and to Puerto Rico. The privately-owned airline began services over a St Croix-San Juan route in February 1980. Later that year, Eastern Caribbean Airways was absorbed. Coral Air entered bankruptcy in 1981, re-emerging a year later under new ownership, operating new aircraft in revised company colours. In 1984 the airline again ran into financial difficulties and underwent a change of ownership. Routes served today include St Thomas-San Juan and St Croix-St Thomas. *(Short Brothers Plc)*

Crown Airways (AL)

Crown Airways, the fifth member of the Allegheny Commuter System, took delivery of a Shorts 330 on 9 September 1980. The company maintains a scheduled route network, connecting points in Pennsylvania, Ohio and West Virginia. The carrier is a subsidiary of the Brockway Corporation, which also owns the regional airline, Brockway Air.

Crown Airways has been a Commuter member since 1969, when flights began between DuBois and Pittsburgh. Routes served by the company include DuBois, Franklin and Pittsburgh in Pennsylvania, Clarksburg and Parkersburg in West Virginia and Youngstown in Ohio. The company employs over 90 people and has its base at DuBois Jefferson County Airport. *(Aerogem)*

Dash Air (WI)

Dash Air ceased operations and filed for bankruptcy in September 1984. The privately-owned carrier had commenced services in June of the previous year, having acquired the commuter airline Air Irvine through a merger of the two companies. Dash Air had quickly developed its route network in southern and central California to serve the cities of Fresno, Los Angeles, Montercy, Ontario, Orange County (Santa Ana), Palm Springs, San Francisco and Santa Barbara. The airline had its base at Orange County's John Wayne Airport and utilized Embraer 110 Bandeirante, Shorts 360 and Piper Navajo Chieftain aircraft. At the time that Dash Air ceased operations, the company had over 140 employees. *(Udo and Birgit Schaefer Collection)*

DLT — Deutsche Luftverkehrsellschaft
(DW)

DLT is a fast-growing West German regional airline formed in 1974, providing scheduled passenger services between various domestic points as well as operating scheduled internal and international flights on behalf of Lufthansa, of which it is a subsidiary. In the late 1970s the carrier was known for a time as DLT — German Domestic Airlines. The majority of the company's shareholding is controlled by AGIV — Aktiengesellschaft fur Industrie und Verkehrswesen, who own 74 per cent, with Lufthansa owning a minority interest of 26 per cent. Although the company no longer does so, DLT used to operate Shorts 330s, one of which, D-CBVK, is seen here awaiting passengers. *(Short Brothers Plc)*

Eastern Airways (EN)

Eastern Airways was the joint operating name of Leasair of Humberside, and the Norwich-based air taxi company Iceni Aviation. The company operated air taxi, passenger and cargo charters throughout Europe and North Africa, as well as various scheduled night mail services on behalf of the Post Office. The company's Shorts 330s operated a five nights-a-week London (Stansted)-Liverpool-London (Gatwick) service and a Saturday service carrying newspapers on the London (Gatwick)-Dusseldorf route on behalf of the British Forces. A twice-daily scheduled passenger route was also operated between Humberside-Norwich-London (Heathrow). This service was initially flown by Douglas DC-3s before being taken over by Shorts 330s, which also flew the route from Humberside to Glasgow when this was established. The company's controlling interest was purchased in 1982 by General Relays, which became part of the new Genair.

(Short Brothers Plc)

Above: **Fairflight** (FC)

Fairflight Ltd is a British charter passenger and cargo company which specializes in contract services in the UK, Europe and North Africa. The privately-owned company was formed in 1969 and has a base at Biggin Hill in Kent. As well as being a major lessor of aircraft — dry, wet or anything in between — Fairflight have a major maintenance and overhaul facility, with a staff of over 80 people, and are capable of carrying out all checks up to major on Shorts aircraft. The regional commuter operator Air Ecosse is a subsidiary company of Fairflight Ltd. The major part of the Kentish-based company's aircraft comprises Shorts 330s and 360s, with Embraer Bandeirante, Twin Otter and Cessna Citation executive aircraft making up the rest of the fleet. This Shorts 360 was photographed at London (Heathrow) while on a service from Carlisle and Dundee. *(P Hornfeck)*

Opposite: **Fischer Brothers Aviation** (AL)

Fischer Brothers Aviation commenced operations in 1951 from its base in Galion, Ohio. It now flies scheduled passenger services as part of the large Allegheney Commuter systems, and links cities in Ohio, Michigan and Pennsylvania. In 1969 the company became one of the first carriers to join the Commuter network. Today it operates a fleet mainly formed of Shorts 330s and 360s, but it also flies Casa 212 Aviocars and a single Piper Aztec. FBA's route network includes Detroit-Cleveland, Columbus-Flint-Kalamazoo, Columbus-Cleveland, Mansfield-Pittsburgh. *(Andy Clancey Collection)*

Opposite: **Genair** (GN)

Genair, one of Britain's largest commuter air carriers, suspended operations and entered receivership in July 1984. The company had begun scheduled services in August 1981 over the Liverpool-London (Gatwick) route. During late 1982 operations were expanded greatly with the absorption of Eastern Airways, which had been in existence since 1972, and the take-over of commuter services of Casair. Genair now had a substantial fleet of aircraft, the majority of which were Shorts 330 and 360 aircraft. In conjunction with British Caledonian Airways, Genair pioneered the Gatwick-based carrier's Commuter Services to link and

"feed" passengers from BCAL's international services to points within the UK. All of Genair's aircraft were repainted into the BCAL colour scheme, including this example, Shorts 360 G-BKKT. Up until the suspension of services, routes flown included Gatwick-Humberside, Leeds/Bradford, Liverpool, Norwich and Teesside.
(Short Brothers Plc)

Above: **Golden West Airlines** (GW)

Golden West Airlines was formed in 1969 following a merger of four commuter carriers: Aero Commuter, Catalina Airlines, Cable Command Airlines and Skymark Airlines. The company provided high-

frequency services within southern California. Routes served included Los Angeles to Santa Ana, Santa Barbara, Ontario, Oxnard, Sandiego, Kiern County. Also included in the route system were Inyokern, Edwards Air Force Base, Palm Dale and San Diego from Santa Ana. Charter flights were also carried out and the company was given a contract to fly mail over these routes, as well as to Avalon Bay, Catalina Island. Golden West filed for bankruptcy on 22 April 1983. Until operations ceased the company operated a mixed fleet of DHC-6 Twin Otters and Shorts 330 aircraft.
(Andy Clancey Collection)

Guernsey Airlines (GE)

Guernsey Airlines flies scheduled passenger services between the Channel Islands and points in England and Scotland, and also undertakes inclusive tour charters with a Shorts 330. The airline was established as a subsidiary of Alidair Ltd in November 1978. Scheduled services were inaugurated between Guernsey and Manchester on 1 April 1980 and a scheduled Guernsey-London (Gatwick) service began three years later. In August 1983 the company was acquired from Alidair by Jadepoint Ltd, who own British Air Ferries. Guernsey Airlines provides flight connections at Gatwick with British Caledonian Airways as a member of the British Caledonian Commuter Group. (Andy Clancey)

Hawaiian Air (HA)

Hawaiian Air was established in 1929 as Inter-Island Airways Ltd, and began scheduled flights between Honolulu and Hilo in November of that year using Sikorsky S-38 amphibious aircraft. Twelve years later, the airline changed its name to Hawaiian Airlines Ltd, and Douglas DC-3 equipment was introduced. Hawaiian Air schedules well over 130 daily flight departures from six Hawaiian Islands. The carrier also undertakes tour group charters and Hawaiian scenic air tours. In terms of passenger traffic, the airline is the largest of Hawaii's inter-island carriers. Hawaiian Air no longer operates Shorts 330 aircraft on their inter-island services, the De Havilland Canada DHC-7 having taken over. *(Aerogem)*

Above: Henson Airlines (PI)

Henson Airlines was established in 1931 as a fixed-base operation. Regular passenger flights began in 1962 under the name of Hagerstown Commuter, linking Hagerstown with Washington DC. Five years later, Henson became the first airline to join the Allegheny Commuter system and remained with them until 1983, when the company moved over to Piedmont, who by 1987 were to have acquired full ownership via its parent company, Piedmont Aviation. Henson flies Shorts 330 aircraft to many points which include Allentown, Boston, Harrisburg, Philadelphia, Richmond and Washington DC. The company's primary traffic centre is Baltimore, and in 1981 it moved its base from Hagerstown to Salisbury, Maryland. *(J Lezark)*

Opposite: Inter City Airlines (QT)

Inter City Airlines, the trading name for Alidair Ltd, was formed in 1972. The company operated scheduled passenger services, long and short term charter and leasing contracts, as well as contract maintenance for many other carriers. ICA's scheduled route network covered a system of services that included East Midlands-Edinburgh on a twice-weekly basis, East Midlands-Aberdeen, and East Midlands-Brussels, both routes being serviced twice daily. The company ceased operations on 1 August 1983. One of the airline's aircraft, a Shorts 330, is seen here en route to Edinburgh. *(Short Brothers Plc)*

Below: **Jersey European Airways** (JY)
Jersey European Airways is a scheduled commuter airline that operates flights within Britain and to France, as well as undertaking charters. The company was formed on 1 November 1979 to take over the services of Intra Airways, a charter airline which commenced operations in 1969 and started scheduled services two years later. JEA is owned by C Walter and Sons Ltd, who also own Spacegrand Aviation. From the company's base on Jersey, in the Channel Islands, the airline flies a route network including Jersey-Dinard, Jersey-Paris, and Jersey-Stansted utilizing Shorts 330 aircraft. There is also an additional service between London (Gatwick) and Dinard. *(Andy Clancey Collection)*

Opposite: **LAPA — Lineas Aéreas Privadas Argentinas** (MJ)
The Buenos Aires-based domestic carrier Lineas Aéreas Privadas Argentinas placed the first of its 330s on its extensive route network during November 1980. The Shorts aircraft, in their somewhat colourful guises, were put on routes from the airline's headquarters at Buenos Aires Aeroparque, and principal domestic destinations include Córdoba, Bahía Blanca, Mar del Plata, Rosario and Tandil, while an international route to Montevideo in neighbouring Uruguay is also provided. Other towns served by the aircraft include Calonia, General Pica, Gualeguaychu, Junín, La Plata, Mercedes-Corrientes, Necachea, Paso de los Libres, Pehujo, Santa Tessita and Cilla Gesell. G-BIFG is seen here prior to its delivery as LV-OJG. *(Short Brothers Plc)*

Loganair (LC)

Loganair was formed in 1962 by Duncan Logan Contractors. In addition to the company's comprehensive passenger network within the mainland and islands of Scotland, services are also provided to points in England, Northern Ireland and the Isle of Man. The carrier also offers charter and contract services throughout Britain, while maintaining the vital Scotland Air Ambulance Service. In November 1983 British Midland Airways acquired a controlling 75 per cent interest in the company from the Royal Bank of Scotland, which had previously owned Loganair since 1968. Loganair flies to many remote destinations, not the least being the Isle of Barra, where services are subject to the tide, as flights operate from the beach, the only suitable air strip. The company also operates the world's shortest air route between Westray and Pappa Westray in the Orkney Islands, flying time being two minutes. The Loganair fleet includes Shorts 360 aircraft, one of which, G-BLGB, is the subject of this study at Manchester. *(P Huxford)*

Malaysian Air Charter (DP)

Malaysian Air Charter of Kuala Lumpur became the first airline in South-East Asia to acquire the Shorts 360. The aircraft was delivered to Kuala Lumpur in September 1984 and it is used on a scheduled service network in Peninsular Malaysia, operating under the name "Macair". Routes flown by this aircraft include Malacca-Kuala Lumpur, Malacca-Singapore, and Kuala Lumpur to the oil town of Kerteh, on the east coast of the peninsula. MAC, which was founded in 1962, is a wholly Malaysian-owned concern. The company was not a first-time Shorts operator since it also owns Skyvans and is the world's largest civilian operator of this aircraft type. Macair operates another route, Singapore-Pulau Tioman, in conjunction with Tradewinds, the charter subsidiary of Singapore Airlines, the former using Macair's 360 and the latter a Tradewinds Skyvan. (Short Brothers Plc)

Metro Airlines (FH & HY)

Metro Airlines commenced operations in 1967 as Houston Metro Airlines, with preparations for flight operations being made by the construction of a commuter STOL airport at Clear Lake City, Texas. The company is a growing regional passenger carrier, now operating extensive scheduled services in Texas, Oklahoma and the South. A publicly-owned airline, Metro has its headquarters in Houston and maintains flights in Oklahoma from an operational centre at Oklahoma City. Via a subsidiary known as Eastern Metro Express, scheduled services with amongst other types, Shorts 330 aircraft, are provided in conjunction with Eastern Airlines from hubs at Atlanta, Houston and San Antonio. Through an arrangement with American Airlines, scheduled flights are operated from Dallas Fort Worth Regional Airport, as part of the American Eagle Commuter system.
(Andy Clancey Collection)

Metropolitan Airways (RD)

Metropolitan Airways, until its demise in 1986, was a scheduled passenger operator, flying services within the United Kingdom. The company maintained certain flights in conjunction with Dan-Air and British Caledonian Airways. Metropolitan had begun operations in 1978 as Alderney Air Ferries, using Islander aircraft to link its base in the Channel Islands with Bournemouth. In 1981 the airline transferred its base to Bournemouth Hurn, and subsequently began commuter flights in conjunction with Dan-Air using Twin Otter equipment. Due to increases in traffic, a Shorts 330 was purchased in 1984. Points served on the company's network included Birmingham, Bournemouth, Bristol and Cardiff. In association with Dan-Air, the company flew a Bournemouth-Cardiff-Manchester-Newcastle-Glasgow route. A seasonal service, also flown on behalf of Dan-Air, linked Bristol with Cork and in addition a Newcastle-Glasgow sector was maintained for BCAL. (Short Brothers Plc)

Above: **Mid-South Commuter Airlines** (VL)

Mid-South Commuter Airlines, formed in 1975 as Resort Commuter Airlines, was absorbed into Air Virginia during 1984. Until the merger, the airline maintained scheduled passenger flights with Embraer Bandeirante and Shorts 330 aircraft. From their base at Pinehurst, services radiated to Charlotte, New Bern, Raleigh and Rocky Mount in North Carolina, Danville, Newport News and Richmond in Virginia, and National and Dulles airports at Washington DC. *(J Lezark)*

Opposite: **Mississippi Valley Airlines** (XV)

Mississippi Valley Airlines was formed in 1969, but it has only really become a force on the US air transport scene since late 1978, when the deregulation process began in earnest. In early 1981 the company was in a very poor way financially. However, cancelling an order for three Fokker F-27s and standardizing on Shorts aircraft, the company began to get back onto its feet. At one stage, with a fleet of twelve 330s and eight 360s, MVA was able to boast that it had the largest fleet of

Shorts aircraft in the world. The first 360 was delivered at the end of January 1983 and entered service on the Quad Cities (Moline)-St Louis leg of the new St Louis-Cedar Rapids route. Three more 360s were delivered during the same year and another four in 1984. MVA was merged with Air Wisconsin during 1985 but its route system has been kept, with cities from Minneapolis St Paul to Kansas City being served. *(K Armes)*

Opposite: **Murray Valley Airlines** (FB)
Murray Valley Airlines moved into a new era in 1983 when it took delivery of its first "wide-bodied" Shorts 330 aircraft. Prior to this the company had been operating an Embraer Bandeirante on its services but it was decided that a larger type was required to meet the growing demand for passengers and freight. With a base at Mildura, MVA flies to Melbourne in Victoria, Broken Hill in New South Wales, and to Adelaide, Renmark and Mt Gambier in South Australia. With the intro-duction of the Shorts aircraft, service frequencies were increased. This additional capacity was well utilized, so much so that two larger 360 types were ordered and introduced onto these services, with the Bandeirante being used as a back-up.
(Short Brothers Plc)

Above: **Newair** (NC)
Newair took delivery of its sole 360 in 1983. The company began operations as a fixed-base air taxi under the name of New Haven Airways in 1962 and only commenced commuter operations some 16 years later, with the Newair title being adopted in 1980. Today the company is expanding rapidly, serving five states. Emphasis lies in linking the Connecticut coastal cities of New Haven, New London and Bridgeport with the New York City, Philadelphia and Washington Metropolitan areas. Typical routings include New London-New Haven-La Guardia, New Haven-Kennedy-Philadelphia and New London-New Haven-Islip-Baltimore. *(Short Brothers Plc)*

41

Okada Air (OX)

Okada Air is a privately-owned Nigerian airline operating domestic and international charter passenger and cargo flights. The company was established in 1983 by Chief Igbinedien, the Esama of Benin City, capital of the southern Nigerian state of Bendel. Shorts 330 aircraft are used on commuter services operating under the title of "Niger Express". *(P Hornfeck)*

Olympic Aviation (OA)

Olympic Aviation, subsidiary company of the government-owned Greek national airline Olympic Airways, began services with Shorts 330s in 1980. The company was no stranger to the manufacturer as it had been using the smaller variant, the Skyvan, since 1970, although the aircraft at that time were used by the parent company as Olympic Aviation was not formed until 1971. The company flies an extensive network of routes covering the regional centres of Greece and the islands in the Aegean Sea. *(Short Brothers Plc)*

43

Above: **Pennsylvania Airlines** (AL)
Pennsylvania Airlines commenced operations in 1965 from its base at Harrisburg International Airport, Middletown. The company became another of the many carriers to join the Allegheny Commuter System in 1973, and was known until 1980 as Pennsylvania Commuter Airlines. Today the airline maintains a route network covering the states of Pennsylvania, New York, New Jersey and Washington DC. The majority of the carrier's fleet consists of 330s and 360s but Allegheney Commuter also operates a King Air and three Twin Otters. There were four Mohawk 298s amongst the

fleet but these have now been withdrawn from use. There were in 1986 daily or weekday flights connecting Harrisburg with New York (La Guardia), New York (JFK), Newark, Philadelphia, State College and Washington. *(J Lezark)*

Opposite: **Royal Thai Army**
In 1984, the Royal Thai Army ordered two Shorts 330/UTT aircraft. The UTT (Utility Tactical Transport) is derived from the passenger version of the 330, but a number of developments have been made in this variant to render it more suitable for its military role. In place of conventional passenger

doors, split inward opening doors are fitted. These can be opened in flight to allow paratrooping, with up to 30 paratroops and a jumpmaster seated in wall-folding seats, and supply dropping. The aircraft can be quickly converted to the medical evacuation role, when 15 stretcher cases and four seated personnel can be accommodated. As an ordnance/freight transport the aircraft can carry a payload of up to 3237 kg (8000 lb), with a roller conveyor system which is capable of handling bulky and heavy goods. *(Short Brothers Plc)*

Royale Airlines (OQ)

Royal Airlines began operations in April 1970, flying between the towns of Shreveport, Fort Polk, Lafayette and New Jersey. By 1986 the company had rapidly grown into a large regional publicly-owned commuter airline, serving 15 points in Louisiana, Texas, Florida, Mississippi and Tennessee from a base at Shreveport, Lousiana, where it employed nearly 500 people. Since September 1983, the airline has operated Houston flights as part of the "Continental Commuter" network in conjunction with Continental Airlines. Both companies coordinate connecting flights at Houston Intercontinental Airport. Royale no longer operate the Shorts 330, having replaced it with Bandeirante and Gulfstream 1 types. *(Short Brothers Plc)*

Shorts SD-330

Following a worldwide market survey in April 1972 to judge carrier's requirements for a commuter airliner, the SD-330 was given a programme launch in May 1973. July of that year saw the metal being cut for the first prototype, with Pratt and Whitney PT6A-45 engines having their initial testbed runs the same month. A month later the first test runs of the five-bladed hush propeller commenced. In April 1974 the first prototype structure was completed, along with delivery of the first pair of engines. Rollout took place 17 July 1974 and a month later the first flight was made. A little over a year later the aircraft was awarded its Certificate of Airworthiness, with initial customer delivery in 1976. The SD-330 seen here is used by Shorts as a demonstration and development aircraft.
(Short Brothers Plc)

Opposite: **Shorts 360**

In the fourth quarter of 1980 Short Brothers announced their intention to introduce a "companion" to its successful 330. Offering 36 seats, the aptly named Shorts 360 first flew in 1982. Unlike its sister the 330 has a single tail unit, and offers an increase in fuel efficiency. The first airline to order the aircraft was Suburban Airlines. The larger capacity offered the airline industry what it needed, especially those companies operating the 330. Engine type chosen was the popular Pratt and Whitney PT6. A new, upgraded variant, the PT6A-65R, was later announced to provide the extra efficiency that the 360 was meant to offer. Seen here is the prototype aircraft, also used for demonstration and development purposes.
(Short Brothers Plc)

Below: **Simmons Airlines** (MQ)

Operators of the world's largest fleet of Shorts 360s, Simmons Airlines of Marquette, Michigan began operations in 1978, initially serving the Marquette-Lansing route. In 1986 the company was one of the fastest-expanding regional carriers in the United States. Services extended to 23 cities in Michigan, Illinois, Wisconsin and Ohio and the airline flew to more cities in Michigan than any other carrier. The first 360s were delivered in 1983 and repeat orders followed with the introduction of more routes. In 1985 the carrier announced that it had entered a joint marketing and scheduling relationship with republic Airlines (under which Republic acquired 9.3 per cent of Simmons' stock) and on April 20 it began flying Republic Express services to Detroit Wayne International Airport.
(Aerogem)

Opposite: **Spacegrand Aviation** (XF)

Spacegrand, along with its sister airline, Jersey European Airways, are wholly-owned subsidiaries of the large Walkersteel Group of Companies and fly a growing scheduled route network throughout the UK, Channel Islands and Northern France. The first 330 was acquired to serve the Paris (Charles De Gaulle)-Channel Islands route as well as the London (Stansted)-Channel Islands service. As a result traffic increased on both routes, even in the depressed winter period. It soon became apparent that both the Stansted and Charles De Gaulle routes required a dedicated aircraft, so another 330 joined the fleet. Concurrently Spacegrand acquired a new route, Belfast-Teesside, and required extra capacity on its routes from Blackpool to Belfast Harbour and the Isle of Man. A third example therefore joined the fleet in April 1985. *(Short Brothers Plc)*

Above: **Suburban Airlines** (AL)

In 1958 Suburban was established as Reading Airlines. It has since graduated from a single Twin Beech, through De Havilland Doves, Twin Otters and a Beech 99 to its present fleet of 330s. Suburban's decision to switch to the "wide body" Shorts aircraft was prompted by the lack of capacity of the Twin Otter. The first 330 went into operation in January 1979, boosting local factors with a degree of passenger acceptance substantially higher than anticipated. By December of the following year, six aircraft had been delivered. Suburban flies a network covering Buffalo, Binghampton and New York City (JFK) in New York, Newark in New Jersey, Allentown, Philadelphia, Lancaster and Reading in Pennsylvania. All services are operated under the Allegheny Commuter banner, a subsidiary of US Air. *(J Lezark)*

Sunbelt Airlines (IM)

Sunbelt Airlines suspended operations in October 1984 due to financial difficulties. The airline had commenced services in February 1979 between Camden, Memphis and Dallas, and until March 1982 was known as Jamaire. Sunbelt's aircraft, including its Shorts 360s, had been operating commuter passenger flights to Camden, El Dorado, Fayetteville and Fort Smith in Arkansas, Jackson, Laurel/Hattiesburg, Meridian and Tupelo in Mississippi, Memphis in Tennessee, Dallas in Texas, New Orleans in Louisiana, and Muscle Shoals in Alabama. *(Short Brothers Plc)*

(J Lezark)

Sunbird Airlines (ED)

Sunbird Airlines began operations in November 1979, with an initial link in North Carolina over a Hickory-Charlotte-Raleigh-Rocky Mount line. The company now maintains regular scheduled Shorts 330 passenger routes which extend to South Carolina, Georgia and Tennessee. In June 1983 a merger between Sunbird and Atlanta Express Airline Corporation took place. A year later, through a change of ownership, financial control of the airline was taken over by different private investors, with Air Transportation Holding Company (Air T) having a minority interest. Air T also owns Mountain Air Cargo. *(J Lezark)*

Sunstate Airlines (OF)

Sunstate Airlines was formed in 1980 by truck dealer Bevan Whitaker. Today the company provides scheduled services using a variety of aircraft including the Shorts 360 over routes in southeast Queensland from its main base at Brisbane. On 1 January 1984 the company fully integrated into its system the airline Noosa Air, which had also been set up by the same owner, although separate identities had been retained. The latter company was formed in 1975 to fly from its purpose-built airport at Noosa to Brisbane International Airport, the carrier using Ansett Airlines' terminal. Sunstate, which also flew from Brisbane, initially operated to Maryborough, Bunaberg, Gladstone and Toowamba. Today, the airline serves Brisbane, Bundaberg, Gladstone, Great Keppel Island, Hervey Bay, Maroochydore, Maryborough, Noosa, Rockhampton and Toowoomba. *(S Wills)*

Syd Aero (UF)

Tuesday 12 June 1984 marked the start of Shorts 330 operations in Sweden when Syd Aero inaugurated its new service from Oskarshamn to Stockholm via Norrköping. The company had commenced services in 1967 from its base at Oskarshamn to points within Sweden. Two Twin Otters were added to the fleet and routes were expanded to cover a variety of towns in Denmark. Following an upsurge in passenger traffic, however, capacity became increasingly short and it was therefore decided to expand the fleet with a larger type. The Shorts 330 was finally chosen and was delivered in 1984. Today, Syd Aero serves Aarhus, Billund, Copenhagen, Esjberg, Odense, Ronne, Skydstrup, Stavanger and Thisted. The aircraft is seen here at Belfast's Harbour airport prior to its delivery as SE-INZ. (Short Brothers Plc)

Above: **Thai Airways** (TH)

Thai Airways purchased four Shorts 330s in 1982 to replace British Aerospace 748s on routes which were no longer economically viable with the older aircraft. The introduction of the 330s resulted in a dramatic drop in operating costs and it was therefore possible to offer increased frequencies and better connections. By August 1983 Thai was able to report that traffic on some routes had grown by over 50 per cent from the start-up of 330 services. It was therefore decided to augment these aircraft by the acquisition of Shorts 360s. Typical routes served by the Shorts aircraft include Phuket-Surat Thani, Hat Yai-Narathiwat, Bangkok-Khon Kaen, Chiang Mai-Nan, Phrae-Lampang. Thai Airways are based at Bangkok. *(Short Brothers Plc)*

Opposite: **Time Air** (KI)

Time Air of Lethbridge, Alberta were Canada's launch customer for the Shorts 360, which entered service during October 1985 on the sector from Vancouver City to three locations on Vancouver Island, Campbell River, Comox and Victoria. In introducing the 360 to this route, Time Air replaced jet services flown by a previous carrier. Prior to the delivery of this type, the airline operated the Shorts 330, having introduced three to its network in 1976. Founded by Walter R Ross in 1966, Time Air has steadily expanded and is now firmly established as Western Canada's major commuter airline, serving 24 communities in three provinces. In addition to the 330s and 360s, the company has operated De Havilland DHC-7 Dash 7s and Convair 640s. *(Short Brothers Plc)*

Westair Commuter Airlines (OE)

Another airline using the Shorts 360 is Westair, a commuter operator which operates over a wide network covering northern and central Califorhia. The company was formed as a successor to Stol Air, another commuter airline formed in 1972, and is a subsidiary of Western Holdings, which in 1983 acquired the airline from the now defunct Pacific Express. The carrier has its main operating base at San Francisco International Airport and operates a route system which includes Chico-Sacremento-San Francisco, Eureka-San Francisco, Crescent City-Eureka-Sacremento-San Francisco, Chico-Sacremento-San Jose-Fresno. The airline is steadily increasing its network and by 1986 employed over 200 people.
(Udo & Birgit Schaefer Collection)

Wings Australia

In late 1980 Jet Charter Airlines (which in February 1982 became Wings Australia) of Sydney won a contract with Queensland Mines to support a uranium mine in Arnhemland using a Shorts 330 based in Darwin. The operation began in January of the following year, after arrival of the first of the four aircraft ordered. The 330, registered VH-KNN, entered service immediately upon its arrival in the Northern Territory, flying between Darwin and Nabarlek, east of Darwin. During the 18 months of operation, the aircraft flew a total of 37 483 hours and was subsequently retired, having had a 98.6 per cent despatch reliability. Three other aircraft were then put onto this route and were operating successfully until the airline suspended operations in November 1984 and was taken over by Pal Air Aviation, who still operate the latter 330s.
(Andy Clancey Collection)

Air UK (UK)

Air UK first became a "wide-bodied" Shorts operator when it placed one 330 into service in early 1983 on its three-times-daily scheduled service between Stansted and Amsterdam. Another aircraft quickly joined its sister, flying the company's Southampton to Rotterdam route. Air UK is one of Britain's largest independent operators and has an operational fleet of over twenty aircraft, six of which are Shorts types, although there were in 1986 plans to purchase or lease more of the 330 and 360 models. *(Andy Clancey Collection)*

British Midland Airways (BD)

British Midland Airways leased a Shorts 330 from the new defunct Inter City Airlines as a stopgap while the company was awaiting its second 360. The aircraft flew a five-a-day service in both directions between East Midlands and London (Heathrow). The performance of the 330 on this route proved very satisfactory to BMA, and although it was slightly smaller than what was needed, its fuel efficiency and cost-effectiveness outweighed this inconvenience. The aircraft was leased by the company in early 1983 and was returned over a year later. *(P Hornfeck)*

61

Loganair (LC)

The first United Kingdom carrier to adopt the "Shorts look" was Loganair of Scotland, which acquired its initial 330 in July 1979 and later added a second 30-seater for operation on a route network covering Edinburgh, Glasgow, Blackpool, Manchester and Belfast. The airline took delivery of one of the larger 360s in March 1983 for use on its Edinburgh-Belfast and Edinburgh-Manchester services. In 1986 the Loganair 360s besides operating on these routes were flying Glasgow-Londonderry, Belfast-Manchester, Manchester-Glasgow, and Edinburgh-Wick-Kirkwall. *(Short Brothers Plc)*

Above: **Manx Airlines** (JE)

Manx Airlines first introduced "wide-bodied" Shorts 330 aircraft into its fleet in May 1983, when it placed the type onto its routes from Ronaldsway, in the Isle of Man, to Glasgow, Edinburgh, Blackpool and Belfast. The 330 was replaced by its larger sister, the 360, due to lack of capacity. Manx by 1986 owned or leased a total of six of the latter type, although due to the increase in routes and passengers carried, further Shorts aircraft were likely to be obtained. *(Andy Clancey Collection)*

Overleaf: **Thai Airways** (TH)

At a ceremony at Short Brothers' Belfast headquarters on 22 October 1985, Thai Airways accepted the first of two 360 Advanced wide-bodied aircraft. The company had evaluated other manufacturer's aircraft but eventually selected the Shorts type since it offered, amongst other things, a significantly lower operating cost over the airline's sectors of up to 250 nautical miles. The 360s were almost immediately put into service to expand operations in southern Thailand, with new routes being introduced to Loei and Saken Nakhon in the north-east and Nakhu Si Thammarat in the south. *(Short Brothers Plc)*